John Codman, David Ames Wells

The Question of Ships

The Decay of our Ocean Mercantile Marine its Cause and Cure

John Codman, David Ames Wells

The Question of Ships
The Decay of our Ocean Mercantile Marine its Cause and Cure

ISBN/EAN: 9783337319793

Printed in Europe, USA, Canada, Australia, Japan

Cover: Foto ©berggeist007 / pixelio.de

More available books at **www.hansebooks.com**

QUESTIONS OF THE DAY—LXIV.

THE QUESTION OF SHIPS

I.—THE DECAY OF OUR OCEAN MERCANTILE MARINE—ITS CAUSE AND CURE

BY DAVID A. WELLS

II.—SHIPPING SUBSIDIES AND BOUNTIES

BY CAPTAIN JOHN CODMAN

G. P. PUTNAM'S SONS

NEW YORK   LONDON
27 WEST TWENTY-THIRD ST.   27 KING WILLIAM ST., STRAND
The Knickerbocker Press
1890

Press of
G. P. Putnam's Sons
New York

THE DECAY OF OUR OCEAN MERCANTILE MARINE: ITS CAUSE AND CURE.

WILL SUBSIDIES BRING BACK OUR SHIPS?

The time has come, when the cause and cure of the extraordinary decline in the ocean mercantile marine of the United States ought to be no longer a matter of doubt and controversy; for the experiences involved have been so thoroughly investigated, are so unquestionable, and admit of such clear presentation, that there is nothing to prevent any citizen of ordinary intelligence from readily understanding the whole situation and arriving at definite and satisfactory conclusions, without resorting in the least degree to hypothesis. As it is evident, however, that there is not as yet a satisfactory understanding and agreement of opinion in respect to this subject, on the part of the American public, it is expedient to present anew the record of our experience and the salient points of the present situation. These are in the main as follows:

PERIOD OF GROWTH AND PROSPERITY OF THE OCEAN MERCANTILE MARINE OF THE UNITED STATES.

So long as the people of the United States controlled the best and cheapest material (wood) for the construction of ships, and through long experience had become most skilful in their building and navigation, so long did the American mercantile marine continue to increase and prosper, even in spite of many disadvantages, in a most wonderful manner.

I   I

The aggregate tonnage belonging to the United States in 1861 was but very little smaller than that of Great Britain, and nearly as large as the entire tonnage of all maritime nations combined, with the single exception of Great Britain.   American-built and -manned ships in 1856 not only carried more than 75 per cent. of all the things (imports and exports) that came in and went out of the country, but more than 50 per cent. of the tonnage of the United States was exclusively in foreign employ,—carrying cargoes, at large profit, from foreign ports to foreign ports, for foreigners, to be used by foreigners,—and in which business Americans had no direct interest but to receive freight money, to be sent home and added to the productive capital of the country.

PERIOD OF DECLINE AND DECAY.

Extraordinary as was the growth of the American mercantile marine, its decline and decay have been even more so.

Thus, in 1856, as before stated, American vessels transported 75.2 per cent. of the value of all the goods, wares, and merchandise exported *from*, and imported *into*, the United States.   In 1888 they transported only 13.48 per cent., and this substitution of service is progressing so rapidly, as to portend, at no distant day, the almost entire disappearance of the flag of the United States, as borne by vessels engaged in foreign commerce, from the ocean.   Out of 72,276,000 bushels of grain exported from New York in the year 1881, not one solitary bushel was carried in an American vessel. Between the years 1878 and 1887, the ocean tonnage of the United States declined in a greater ratio than that of any other maritime nation.

CAUSES OF DECAY AND DECLINE IN AMERICAN SHIP-BUILDING AND SHIP-USING.

The decline in American ship-building, and in the American carrying trade upon the ocean, did not, as is often asserted and somewhat widely believed, commence with the war, and was not occasioned by the depredations of the Confederate cruisers. These agencies simply helped on a decadence that had previously commenced; the primary cause of which was the substitution of steam in the place of wind as an agent for ship propulsion, and of iron in the place of wood for ship construction. These substitutions passed from the domain of experiment to that of fact about the year 1837; and the merchants and mechanics of England, speedily recognizing that through these changes the advantages enjoyed by the Americans so long as vessels were built of wood and propelled by sails would be neutralized, with characteristic Anglo-Saxon enterprise, and without any co-operation from government, went speedily to work to make the most of the new conditions, and built, launched, and operated the first ocean steam and iron vessels.

In the first respect, namely, the application of steam to ocean navigation, the Americans were not lacking in shrewdness and enterprise. They waited until English experience had proved the fact to their full satisfaction, and then embraced the idea so eagerly, and turned it to practical account so rapidly, that the foreign steam-tonnage of the United States, which really commenced to exist in 1848, nearly equalled in 1851 the entire steam-tonnage of Great Britain, of longer growth, and continued to increase regularly and largely until 1856. But between 1848 and 1855, the world had acquired some additional information. It had learned that for all practical purposes an iron ship was superior to a wooden ship, and in the long run, cheaper. The immediate result of this was, that the great business of building wooden ships in the United States for sale to foreigners began to decline; falling off from 65,000 tons in 1855, to 42,000 in 1856, 25,000 in 1858, and 17,000 in 1860; so that if the war had not occurred, it was certain that this branch of domestic industry would be substantially destroyed.

Again, although warned of the consequences in the most remarkable and prophetic manner by those most conversant with the situation,* the Americans, nevertheless, continued to use wood exclusively for the construction of vessels. They built their ocean steamships of this material, and they continued to use the paddle-wheel, when England was abandoning it for the screw. The further result was, that the total tonnage of every description built in the United States declined from 583,450 tons in 1855 to 378,804 tons in 1857, and 212,892 in 1860, a reduction in five years of 68 per cent. ; and our ocean steam-tonnage, which in 1851 was nearly equal to that of Great Britain, so dwindled away, that in 1860–61, before the outbreak of the war, there were no ocean steamers, away from our own coast, anywhere on the globe, except perhaps on the route between New York and Havre, where two steamships may have been in commission in 1861, but were soon withdrawn.

Had matters been allowed to take their natural course ; had Americans been allowed simply to take the advantage of the world's progress which was taken by their competitors ; and had not a subsequent restrictive commercial policy made foreign trade to American merchants almost impossible, it is certain that, even in spite of the war, there would have been no permanent material decline in the American shipping interest, and no condition of things to bewail, such as exists at present. But matters were not allowed to take their natural course. The means and appliances for the construction of iron vessels did not then—1855–60—exist in the United States ; while England began to construct iron steamships as far back as 1837. The facilities for the construction of steam-machinery adapted to the most economical propulsion of ocean vessels were also inferior in the United States to those existing in Great

---

* See remarkable letter of Capt. John Codman contributed to the *N. Y. Journal of Commerce* in the spring of 1857, Wells' " Merchant Marine," p. 51.

Britain. Unwise conservatism, antagonizing the adoption of new methods and ideas, contributed in part to this result. A protective tariff on iron, which enhanced the price of this metal in this country to at least 24 per cent. in excess of its average price in Great Britain, obstructed its use and rendered the construction of iron ships and steam-machinery on terms of equal cost with Great Britain an utter impossibility, even had the appliances for their construction been provided. And, finally, a provision of our navigation laws, enacted with a view of protecting American shipping, absolutely prevented citizens of the United States, interested in ocean commerce, from availing themselves of the results of British skill and superiority in the construction of vessels, when such a recourse was the only policy which would have enabled them at the time to hold their position in the ocean-carrying trade in competition with their foreign rivals, and afforded opportunity for adjustment to the new conditions. All other maritime nations found themselves at the same time under the same disabilities as respects the construction of iron vessels as the United States experienced. Neither France, Germany, nor Italy had suitable ship-yards, or the tools and appliances, or the skilled workmen for so doing. But no one of them adopted the policy of the United States. On the contrary, taking a practical common-sense view of the situation, and setting sentiment aside, they concluded it would be the height of folly to permit a great and profitable department of their industries to be impaired or destroyed, rather than allow certain improvements in the management of its details, because suggested and carried out by a foreign nation, to be purchased and adopted. And they, therefore, virtually said to their own people, " If England can build better and cheaper ships for ocean commerce, and will furnish them to you on terms as favorable in every respect as are granted to her own citizens, and if your private judgment and feeling of self-interest

prompt you to buy and use such ships, the state will inter-
pose no objections to your so doing." And the merchants of
these maritime countries, adopting the course which seemed
best to them under the circumstances, went to England and
supplied themselves with ships and steamers of the most
approved patterns, and sharing with England the monopoly
of owning and using the same, have never had any such
results as the United States have experienced; but, on the
contrary, have seen their commercial tonnage and carrying-
trade on the high seas largely increase; and if their shipping
interests have since experienced any vicissitudes, they have
not in any one instance been referred to influences even
remotely connected with the liberal policy that was adopted.

Next in this history came the war, which helped a decadence in our mercantile marine, which, as has been shown,
had already commenced. But the influence of the war after
its termination would have been but temporary on this as
it was on other of our great industries, but for the continuance and extension of a national fiscal and commercial
policy which made it more difficult than ever for an American
merchant to build or use ships as cheaply and effectively as
his foreign competitors, and which also practically destroyed
the business upon which an ocean marine must depend for
profitable employment, or even existence.

THE POLICY OF THE UNITED STATES ANTAGONISTIC TO
FOREIGN COMMERCE.

To appreciate fully the truth of this statement it is necessary to bear in mind, *first*, that foreign commerce is the
exchange of the products of one country for the products
of other or foreign countries; and, *second*, that it is the one
great characteristic feature of the protective policy to restrict or prevent such exchanges. To prove there is no mis-
take in this second proposition, attention is asked to the
following evidence:

The late Henry C. Carey, who stands in relation to the modern doctrine of protection very much the same as Mahomet does to the religion of Islam, expressed the opinion, over and over again, that the interest of the United States—material and moral—would be greatly benefited if the Atlantic could be converted into an impassable ocean of fire ; and also that a prolonged war between the United States and Great Britian would be one of the best possible things for the former country. The late Horace Greeley taught substantially the same doctrine, and in 1872, when a candidate for the presidency, said : " If I could have my way, I would impose a duty of $100 on every ton of pig-iron imported," or, in other words, he would not allow any ship entering a port of the United States to transport any pig-iron into the country. Senator Frye, of Maine, in a speech at a " Home-Market Club " dinner in Boston, October 24, 1888, declared that he wanted " to see duties increased," so that no manufactures of silk or of wool or of iron and steel could be imported. Ask also the protected representatives of all other domestic manufactures, or the producers of raw or crude materials used by manufacturers, and it will be rare to find one who would not agree with Senator Frye in respect to the tariff treatment of his specialties.

Prof. R. E. Thompson, of the University of Pennsylvania, in his " Social Science and National Economy," which is used as a text-book in the university, after devoting some pages to showing the comparative undesirability of foreign trade, expresses his sentiments in regard to it in the following language : " *We have already given some reasons why commerce between distant points is an undesirable thing*" (page 222). " *If there were no other reasons for the policy that seeks to reduce foreign commerce to a minimum, a sufficient one would be found in its effect on the human material it employs. Bentham thought the worst possible use that could be made of a man was to hang him ; a worse still is to make a common sailor of him.*"

Certainly this method of putting an end to the bad influences of foreign commerce would soon reduce it to something less than a minimum, for if we were to hang all the sailors there would be nobody to man our ships, and if our ships could not be manned there would soon be no ships, and without ships so much of foreign commerce as is dependent on ships for ocean transportation would cease to exist; and if from humanitarian motives it was decided not to hang all the sailors, but to compel them to follow other employments less detrimental to their morals and manners —such, for example as working in Pennsylvania coal mines —it would be, according to Prof. Thompson, an economically wise and desirable measure, "for the work of sailors," he says, "while the most difficult of human employments, is also the most unproductive, the most useless."

To put the most favorable interpretation, therefore, on Prof. Thompson's words and teachings, he unmistakably stands upon record as holding the opinion that all foreign commerce is inexpedient, except so far as it can be carried on by land and without the instrumentality of ships, which would necessarily limit the foreign commerce of the United States at the present time to their exchanges between Mexico and Canada, the aggregate of which is comparatively trifling.

What sort of commerce Prof. Thompson would have between the United States and foreign countries he thus sets forth : " *If we take commerce in the largest sense, as meaning the whole intercourse of nation with nation, it will include the interchange of ideas, the naturalization of better political and industrial methods. And with this intellectual exchange there would be associated a commerce in those articles whose artistic excellence and elaboration of workmanship present in a concentrated shape the flower of the nation's intellectual life and spirit.*"

That such a transcendental commerce, such an exchange of bric-à-brac, does not in Prof. Thompson's mind include the great bulk of the foreign commerce of the United States, is clearly shown by the following quotation from a paragraph immediately preceding the sentences last quoted: "Every nation contains within its own providential boundaries the means of making itself independent of all others as regards the supply of articles of prime necessity.* There is, therefore, no need of employing a large number of its people and a large amount of its capital in transporting these articles across the ocean."

It is evident, therefore, that if the economic ideas which Prof. Thompson teaches, and the University of Pennsylvania sanctions, are to prevail, full eighty per cent. of the present export trade of the United States—our agricultural, mining, forest, and fishery products,—which her people and her rulers are now most anxious to extend, would be put an end to, as economically unwise, unnecessary, and unprofitable.

Joseph Wharton, a leading citizen of Philadelphia, and a president of the so-called "American Industrial League," in an article contributed to the *Atlantic Monthly* some years since "On International Trade," adopted as a motto pertinent to his argument, the following words, which Goethe puts into the mouth of Mephistopheles,—or the Devil:

> " *Talk not to me of navigation ;*
> *For war and trade and piracy—*
> *These are a trinity inseparable.*"

---

* This statement is on its face an absurdity. There is not a nation on the face of the globe which has risen above the requirements of a barbarous existence that has "the means of making itself independent of all others as regards the supply of articles of prime necessity," according to the civilized interpretation of terms. Every breakfast table in the land is a protest against Prof. Thompson's assertion. England cannot supply itself with food; Europe with cotton and tobacco ; the United States with sugar, tea, coffee, spices, or dye-stuffs ; Mexico with coal ; and so on. The law of nature, founded on, and an inevitable sequence of, the diversities of race, intellect, climate, and culture, is that man and nations alike everywhere are not independent, but interdependent, and becoming more and more so as civilization increases.

Again, in a debate in the United States House of Representatives, March, 1882, on the features of our existing consular system, the Chairman of the Committee on Appropriations, Representative Hiscock (now Senator), from the great commercial State of New York, admitted that the existing system was "complex," and " an obstruction to the importation of foreign commodities"; and for the latter reason he declared himself in favor of its continuance; for he said: "I am unable to see how, when you relieve the commerce of the country of the weight and burden of our consular system, you are not to that extent abating the protection which is given to our industries."

Abundance of other illustrations to the same effect might be given, but enough of unimpeachable evidence has been offered to prove that the men who for the last quarter of a century have shaped and determined the fiscal and commercial policy of the United States, and are at present in control of the executive and legislative departments of the government, do not believe in international commerce; do not believe in the continuance and enlargement of the business for which alone ships are needed, or in the conditions which alone make the existence of an ocean mercantile marine possible. The men who have adopted these ideas have furthermore not been simply theorists. They have not stopped with mere believing, but having the opportunity, they have embodied their ideas into statutes, and made them the law of the land. And the public officials charged with the administration of the law, taking their cue from the expressed views of the law-makers, seem to regulate their conduct in office on the theory that foreign commerce is an offence which they are in duty bound to discourage; and accordingly, as has been especially exemplified under the present administration, eagerly take advantage of every doubtful point in the wording of the statute, to make a construction on the side of illiberality.

Whatever of decay and disaster has come to our ocean mercantile marine has clearly, therefore, not been the result of accident, but of design,—manifesting itself not in open and avowed hostility to ships, for such a course, on account of national historic associations, would not have been politic, but design in the sense of perfect willingness that our ocean-carrying trade should perish, if thereby the free exchange of the products of the United States for the products of other countries could be restricted or prevented ; and the instrumentalities by which such design has been made reality, are substantially as follows :

First, By the maintenance of a system of navigation laws, which were avowedly modelled on the very statutes of Great Britain which the Americans as colonists found so oppressive that they constituted one of the prime causes of their rebellion against the mother-country,—the main features of difference between the two systems being, that wherever it was possible to make the American laws more rigorous and arbitrary than the British model, the opportunity was not neglected. And these laws, without material change, hold their place to-day upon our national statute-book. International trade since their enactment has come to be carried on by entirely different methods : ships are different ; voyages are different ; crews are different ; men's habits of thought and methods of doing business are different ; but the old, mean, absurd, and arbitrary laws which the last century devised to shackle commerce remain unchanged in the United States, alone of all nations ; and what is most singular of all, it is claimed to be the part of wisdom and the evidence of patriotism to uphold and defend them.

The main provision of these laws is one which forbids an American citizen, if he can buy a vessel cheaper and better suited to his wants in a foreign country, from availing himself of the opportunity. No American citizen is allowed to import a vessel of foreign-build, in the sense of purchasing, acquiring a registry or title to, or using her as his own property,—the only other absolute prohibitions of imports, on the part of the United States, being in respect to counterfeit money and obscene publications or objects. And from this last circumstance the inference is fully warranted that in the eyes of American legislators the importation of a foreign vessel must be prejudicial in the highest degree to the morals of the country.* Note now the effects of this law.

Experience having demonstrated that the ships of the United States cannot do the work which the commerce of the world needs to have done as cheaply and as conveniently as the ships of Great Britain and other competitive maritime nations, the representatives of the world's commerce, who do not mix up business and sentiment, and who simply ask who will serve us best and at the cheapest rates, do not employ American ships ; and for the same reasons, the former great business of building ships in the United States for sale to foreigners no longer exists.

Furthermore, while we are the only people in the world who are forbidden to purchase foreign-built vessels, we freely permit all the world to enter our ports with vessels purchased in any market. Precluded, therefore, by the first provisions of our navigation laws from engaging on equal terms in the carrying trade with foreigners, we wonder and

---

* Although the law (Revised Statutes of the United States, Sec. 4,132 and 4,133) which denies to citizens of the United States registry, protection, or ownership of foreign-built vessels is very clear and explicit, there is reasonable doubt of its constitutionality. The late Caleb Cushing, Attorney-General of the United States, 1853–57, gave an opinion, that a bale of goods, *or any property*, purchased abroad and paid for by an American citizen, became American property, and as such was entitled to the protection of the flag. This opinion was subsequently but unofficially laid before Hon. Amos T. Ackerman, Attorney-General of the United States in 1870, and elicited the opinion, that a vessel purchased by an American citizen in a foreign port, and covered with the American flag, was entitled to her register the same as an American-built vessel.

complain that the carrying-trade of our own products has passed from our control.

Numerous other provisions of our navigation laws have contributed in a lesser degree to the destruction of our ocean mercantile marine ; but it is not proposed to say further of them in this connection, except to call attention to the curious circumstance, that not a single writer or speaker of note, who in recent years has undertaken to defend them, or oppose their repeal, or modification—through lack of knowledge, or more probably a well-grounded apprehension lest a full exposition would of itself defeat his argument,—has ever ventured to tell his hearers or readers, what the code really embraces, or make clear its details.

A *second* instrumentality which has contributed in an even greater degree to the decay and almost absolute destruction of our ocean mercantile marine, has been the enactment and maintenance of laws by the men who have for so many years shaped and controlled our national fiscal and commercial policy, and who, as has been demonstrated, disbelieve in the desirability of foreign commerce, which by the imposition of enormous taxes on imports—amounting in 1887 to an average of 47 per cent. of the value of all dutiable imports, and 31 per cent. on the value of all imports—practically forbid American manufacturers, agriculturists, and merchants, from receiving the products of other nations in exchange or payment for their own ; which say, in fact, to the citizens of Chili and Mexico, " We want to sell you our cotton fabrics and agricultural implements, but you shall not sell us your ores of copper, or of silver-lead "; and to the producers in the Argentine States, Australia and South Africa, " We want to sell you clothing, boots, and shoes, machinery and hardware, but we won't buy the principal product—wool—which you have got to sell—or pay with—in return." But in thus shutting out the products of other nations, we have at the same time necessarily shut ourselves in. For all commerce—foreign and domestic—is simply the exchange of products; so that he who won't buy can't sell, and he who won't sell can't buy.

The attempts to invalidate these conclusions seem almost puerile; but, nevertheless, as they continue to be made by respectable journals, it is expedient to notice them. It is asserted, for example, that it is not necessary to import in order to export. But that is equivalent to saying that a nation can or will go on selling to other nations without receiving pay for what it sells, which ignores the economic axiom, that in the long run the exports and imports of every nation must pay for each other, or the trade will cease,—a fact that would practically appear in every national trade statement, but for the circumstance, that imports, as in the case of England, are often made for the purpose of *paying* interest on foreign investments which represent long antecedent exports; or obligations of the indebtedness (in the shape of bonds) are exported in the place of merchandise to pay for imports, as is often the case with the United States.

It is also asserted that it is not necessary for a country to receive the ordinary products of other countries in payment of its exports, but that payment may be made by an import, or return of the precious metals. The answer to this is, that no nation can spare sufficient of its gold—the standard money of international trade—to pay for even so little as the average value of its importations for a single month; the unexpected export of a million and a half of gold from the United States in September, 1889,—with a net balance of $189,000,000 in the national treasury,—having caused a thrill of disturbance to run through every financial and commercial interest of the country. The first question a representative of any of the states of South America would naturally put when asked to consider a proposition to buy more—*i. e.*, extend trade—from the

United States and pay gold for such increased exports, or purchases, would be: " Where are we to get the supply of gold essential for such a method of trade? and does the United States propose to drain us at once of our gold, and so precipitate a financial panic among our people?"

How successfully the present fiscal and commercial policy of the United States has operated to restrict foreign commerce, or in limiting our markets in foreign countries, is shown by an almost universal recognition of the fact, that for the lack of such markets as our foreign competitive nations possess, our surplus of manufactured products is pressing with smothering effect upon our whole circle of industries. What effect it has in restricting our markets, especially in South America, is shown by the fact, that while 6,607 steamers entered and departed from the ports of the Argentine Republic in 1887, not one bore the flag of the United States; while in 1888 only seven steamers arrived from the United States, and these were all foreign tramp steamers, which go everywhere, upon the shortest notice, in search of freights affording the minimum of remuneration. It is, therefore, clear that it was not from lack of instrumentalities for inter-communication, or cheap rates of freight, but lack of business, that limited the exports of the United States to the Argentine Republic in 1888 to the capacity of seven tramp steamers, averaging in the aggregate less tonnage than one of our great transatlantic steamers. Had the business been possible, not seven, but seventy tramp steamers would have been on hand to compete for it.

A few years after the war, a well-known commercial firm in Boston, which before the war had a large trade with the west coast of South America—and particularly with Chili,— attempted to regain the trade which the war had interrupted. For this purpose they established a line of steamers to run regularly between Boston and Valparaiso. The vessels—

screw steamers—were built in England and owned in the United States, but owing to the provisions of our navigation laws, their registry was in London and they carried the British flag and were commanded by a British captain. So far as the instrumentalities for doing business were concerned, the Boston merchants put themselves on a perfect equality with their foreign competitors. There was no difficulty, moreover, in obtaining full outgoing cargoes, for there was then, as now, a demand in Chili for American productions— cotton fabrics, sewing-machines, woodenware, hardware, machinery, and the like. But ships, to be profitable, must earn freights both going to and returning from a market, and the only commodities which Chili had to give in exchange for our products were copper, copper-ores, and wool, on the importation of all of which the United States imposes wholly, or nearly, prohibitive duties. The consequence was that these Boston steamers, in order to obtain return cargoes from Chili, were obliged to take a freight of wool and copper on English account, and on arrival in Boston, trans-ship it in bond in an English vessel for Liverpool. It is almost needless to say that such a roundabout way of doing business did not pay, that the American line of steamers in question was soon withdrawn, and that since then no citizen of the United States has ventured to repeat the experiment.

One more incident is necessary to complete this story. Some years ago a roving commission, composed of men who had little or no practical or theoretical acquaintance with commerce, was sent by the United States to Central and South America, for the purpose of determining how more intimate commercial relations could be established between these countries and the United States. In due time they came to Chili, and had an audience with its president. They laid before him the purport of their mission, and asked him to consider the negotiation of a treaty establishing a reciprocity of trade between Chili and the United States. The

Chilian president politely but decidedly declined to consider the subject. " It was out of no want of respect," he said, " for the United States; but it was his settled belief that all treaties were needless ; that there could be no control by any convention of the laws of trade ; that men would buy and sell where it was most for their advantage ; and that this could not be aided or materially influenced by national compacts." In conclusion he further remarked that " *Chili opened all her coasts to the vessels of any nation, the United States included,* and in turn the Chilian flag ought to have access to the ports of the United States in like manner." Commenting on this satirical though eminently sensible remark, the United States Commission, in their report of the interview, use this language : " Of course it was not worth while to dwell upon such an avowal." How far the Chilian people were in sympathy of opinion with their president may be inferred from the following comments on the object of the United States Commission in question, which appeared in the leading newspaper and government organ in Valparaiso.

" We believe," it said, "that the United States do not find markets for their products in South America, because the United States has shut her doors to the products of South America. The United States, by means of its heavy tariff, has proposed to realize the impossible, or the selling to all the world without buying any thing from anybody. This being so, it does not need much keenness to discover the origin of the evil and to point out the remedy. If English goods come here in large quantities it is because the ports of Great Britain are open to Chilian products. If we buy of the English, it is because they do not repel through a protective tariff the articles we produce, and of which we can avail ourselves to pay for what we buy ; and if the United States desire to enjoy the benefit which the English reap from this commerce they have only to follow their example—lowering their tariff and opening their ports to us. Such a measure would be much more

efficient for the object sought for by the honorable plenipotenti-
aries than their manifestations of friendly feeling, which, not
being seconded by the practical measures above stated, cannot
produce any favorable change in the conditions of the commerce
of the United States with the people who inhabit this (South
American) continent."

The following is a further illustration of the manner in
which our existing fiscal policy closes the markets of the
world to our surplus manufactured products, and renders
foreign commerce and the maintenance of an ocean
mercantile marine on the part of the United States a
practical impossibility. During the the year 1887, the
United States imposed tariff taxes on the import of per-
fectly crude or raw materials, and on articles wholly or
partially manufactured—all imports for use in the manufactures or mechanic arts of the country—to the extent of
$40,000,000. But *forty* millions is ten per cent. on $400,-
000,000 of product into which these crude materials enter
as constituents, and to such an extent must enhance its
price when offered for sale, if the manufacturer would recoup
himself for its payment. But no business man needs to be
told, that not a dollar's worth of such an immense amount
of product can be sold in any foreign market in competition
with the manufacturers of similar products in Great Britain
and other countries, who are exempt from such a burden of
taxation ; or what is the same thing, who are by our own
acts given an advantage of *ten* per cent., or can undersell
American producers and exporters to that extent in any
neutral markets of the world. And as a matter of fact, we
find that out of $683,860,000 worth of goods, wares, and mer-
chandise, exported from the United States during the fiscal
year 1888, less than 20 per cent. (19.05 exactly) were manufactured articles.*   But great as are the present restrictions

---

* The men most conversant with the practical application of electricity in the
United States assert, that the present difference in the price of copper between

on the foreign commerce of the United States and the almost insuperable obstacles in the way of having an ocean mercantile marine, it is well at this point also to consider what would happen, if Senator Frye, and others who agree with him, could succeed in having the duties at present levied on dutiable imports, increased, as they say they ought to be, to the extent of absolute prohibition. In such a case the United States would receive from foreign countries only such few products as are at present not dutiable, and so not antagonistic to the full operations of the protective policy. This would at once reduce the existing volume of our foreign commerce about two-thirds. We should, on the basis of 1888, exclude foreign imports of merchandise to the extent of $480,000,000. But as imports are made solely for the purpose of obtaining, or paying for exports—product being given for product,—such an exclusion of imports would at once reduce the export of the products of American labor in a corresponding degree. If, under stress of circumstances, foreign nations should so greatly need our cotton, cereals, and other crude products, that they would be willing to pay for them in gold, of what use would such an import of gold be to us? We have already more than $700,000,000 of gold in the country, of which a very considerable part is not in use as currency. We could not eat it, wear it, or use it in any way, except to exchange it for articles of foreign production, and these in turn to be used and enjoyed, would have to be used and enjoyed out of the country. The evidence is thus complete, that so long as we maintain a commercial policy that seeks to restrict or prevent commerce with other nations, so long ships will not come back to us; for opportunity for their

the United States and Europe, due solely to an unnecessary tariff tax on the imports of copper, "is enough to swallow up all the profit on the export of many electrical supplies, such as insulated wire and cables, while it places American makers of dynamos, and of other apparatus containing copper and brass, at a decided disadvantage in the markets of the world."

employment would be limited, even if they were placed as free gifts at our wharves.

Such then is a picture of the situation in which our former great industry of ship-building and ship-using finds itself, not one essential particular of which as has been presented can be fairly questioned or refuted. The expulsion of the Moors and Jews from Spain under Ferdinand and Isabella and their successors, and the revocation of the " Edict of Nantes," which deprived France of her best artisans and industries, have been accepted by all historians and economists as the two most striking and exceptional examples in modern times of great national industrial disaster and decay directly contingent on unwise and stupid, but at the same time deliberately adopted, state policies. It has been reserved for the United States, claiming to be one of the most enlightened and liberal nations of the world, after an experience of near three hundred years since the occurrence of the above precedents, to furnish a third equally striking and parallel example of results contingent on like causes, in the decay and almost annihilation of a great branch of domestic industry, which formerly, in importance, ranked second only to agriculture.

HOW CAN WE BRING BACK OUR SHIPS AND INCREASE OUR FOREIGN COMMERCE?

Having thus exhibited the inception and causes of the decay of our ocean mercantile marine, the way is now clear for a consideration of the methods and feasibility of bringing back ships of the most desirable character, as instrumentalities for the profitable employment of the labor and capital of the United States, and for increasing our foreign commerce, which can alone give employment to an ocean marine and afford a market for the surplus products of our industries.

And first, if the primary cause of the decline of American

shipping employed in the ocean carrying trade was due (as beyond all question it was) to the fact, that American ships could not do the work which the trade and commerce of the world required to have done as cheaply, as expeditiously, and as conveniently as the ships of Great Britain and other competitive maritime nations; if the inception of this decline was coincident with the recognition of this fact by American and foreign merchants, and if the same causes which in the first instance arrested the growth and occasioned the decay of American ocean tonnage have ever since continued and are now fully operative, then it needs no argument to prove that the first step to be taken in the way of recovery is for the American shipping interest to put itself on a par with its foreign competitors in respect to the excellence of the tools or instruments—*i. e.,* the ships and all their appurtenances—which it needs to employ in the transaction of its business. Unless this first step is taken, unless this primary and indispensable result can be effected, there is no use of further talking ; and we might as well fold our hands and complacently say : " We do not propose to be a maritime nation." People in this age of the world will no more continue permanently to use poor or unnecessarily expensive tools in trade and commerce, than they will in agriculture and manufactures. They will either, as the outcome of intelligence, voluntarily adapt themselves to the new conditions that may arise, and so prosper ; or, as the outcome of ignorance and obstinacy, adhere to the old, and be crushed and starved out of existence.

Steamships suitable to meet the present requirements of the commerce of the world cannot be built at the present time in the United States as cheaply as they can be in Great Britain. Steamships of the best quality can be and are being built in the United States for the use of its navy, and for coast and inland navigation. But such ships are not subjected to any foreign competition. The exact extent of this disparity of cost cannot be readily stated. The disparity due to the difference in the comparative prices of iron and steel in the two countries increases the cost of American vessels, in the opinion of experts, to at least fifteen, and probably a greater percentage. The disparity in the cost of machinery is much greater, and greater furthermore at the prevailing low prices than it was twenty years ago, when the actual prices were higher than now in both countries. The best cotton manufacturers in the United States are importing cotton machinery from England, paying duties of 45 per cent. and more on the same, and think they find their advantage in so doing. First-class freighting screw steamers built of steel and thoroughly equipped with all modern appliances were built in Great Britain in 1888 for $34 per ton, as compared with $90 per ton in 1874. But be the disparity of cost between American and foreign-built steamships * greater or less, it is sufficient to recognize in this connection that it is considerable; and to expect that under such circumstances the former can successfully compete with the latter in the same sphere of employment, is as idle as to expect that a man with his feet in a sack can compete in a race with one whose limbs are free and unshackled.

How American business men who still want to do business on the ocean recognize this condition of affairs and adapt themselves to it, is best shown by a few actual examples. Thus, the last report of the Pacific Mail Steamship Co., of New York, announces that there is now building for the company, in Great Britain, a steamer of

---

* It is claimed that steamships of absolutely the first class, like the *Etruria* and *Teutonic*—owing to the greater skill and efficiency of American labor, which, although nominally high-priced, is cheap because of its efficiency—can now be built in the United States on terms as favorable as in any other country; but this claim is not preferred in respect to steamships constructed mainly for freight service.

5,000 tons ; and it is made a matter of congratulation to the stockholders, that this is to be effected at a saving in cost of one third as compared with prices asked by builders in the United States. Some years ago the Pennsylvania Railroad determined to establish a line of transatlantic steamships to run in connection with its road between Philadelphia and Liverpool. At the outset the company proposed to use steamers of American construction only, and did provide itself with four vessels of this character. But *subsequently*, finding itself in need of new steamships, it quietly discarded Pennsylvania's pet theories about American industry and employment of home labor, and supplied its necessities with ships of British construction ; and, with infinite effrontery, *subsequently* memorialized Congress to specially grant to them a registry, in order that this company alone in the United States might enjoy the privilege of lawfully holding such vessels as property.

Wilmington, Delaware, is a place where vessels of the latest style of construction and material can be built. But the merchants of Baltimore, which city is in close proximity to Wilmington, do not go there for their supply of ocean steamers, but use British-built vessels, which they cannot own or run as American property, in their extensive fruit trade with Central America and the West Indies ; and the same is true of those lines of steamers which are engaged in like business running out of Philadelphia.

The following is an even more striking instance of the disadvantage the merchants of the United States labor under in being prohibited from buying foreign vessels for use in their own business. There were, a few months ago, twenty-four Norwegian steamers running on time charters in the fruit trade between New York and Baracoa, and more than half of these steamers were built in Great Britain or other countries foreign to Norway. What does this mean? It means that our stupid system prohibits our merchants from putting their capital into vessels where it could be employed to advantage in bringing goods to our own market, and compels them to employ foreigners to do the business and pocket the freight money.

During the present year there have been under construction in the ship-yards of Great Britain about 200,000 tons of shipping (of iron or steel) for foreign owners. Of this amount Germany is reported to have ordered 80,000 tons; little Portugal, 20,000; France, Norway, and the British Colonies, 10,000; but the United States is credited with only one steamer of 5,000 tons, building for the Pacific Mail S. S. Co., and a bark at Glasgow on Boston account.

Now, to repeal the prohibition against the purchase of foreign ships by citizens of the United States, is not a free-trade measure in the sense in which this term is generally used, but a measure in the largest interest for protection. It is a measure not so much with a view of setting our commerce up, as for removing an obstacle to its setting itself up. It is a genuine American policy according to the doctrine of protection, inasmuch as it will tend to promote and develop a great branch of domestic industry; while the present policy, which pretends to be genuine protection, is really so promotive of European interests that almost every maritime nation is increasing its ocean tonnage at the expense of the United States.

But a measure more essential for the restoration of our shipping and our ocean carrying trade is a radical reform of our whole tariff system and policy. We have got to recognize the fact that it is just this system and policy that has made it impossible to maintain our status as a commercial nation upon the ocean. We have got to recognize that the present pressing necessity of the United States is extended markets for the continually increasing surplus of our products; and that such markets cannot be obtained, or a national commercial marine find a basis for growth, or even existence, so long as we restrict by law the producers of this country from freely exchanging the products of their labor with the products of the labor of the producers of other countries. But these are just the conditions which the representatives of the policy which has driven our ships from the ocean and destroyed our foreign markets do not propose to recognize. They propose not only that no part of our protective system as embodied in our present tariff shall be destroyed, but rather that further restrictions on commerce shall be multiplied. They claim that the present policy can be maintained, and the decadence of American shipping arrested, and an era of maritime prosperity inaugurated, by the payment of shipping subsidies, which are in the nature of bounties. Or, in other words, having almost completely destroyed a great branch of domestic industry, by compelling it to submit to the unnatural restraints of an artificial system, it is now proposed to repair the damage, not by removing the cause, but by resorting to another artificial expedient—namely, the hiring of men to do what the first artificial system makes it for their interest not to do. On its very face could any thing be more economically monstrous and impracticable? But, discarding all matters of sentiment, let us examine this question from a purely practical point of view.

THE QUESTION OF SUBSIDIES.

The first objection to this scheme is, that it is a mere palliative, and even if remedial in part, and unobjectionable as a matter of public policy, bears no proportion to the magnitude of the trouble to be dealt with. Capital and men can be undoubtedly hired to float the American flag. But does anybody suppose that with the present temper of the American people in respect to subsidies and the expenditure of public moneys raised by taxation, that the policy of paying bounties can be indefinitely continued as to both time and amount. But they must be so continued on the bounty theory, unless the causes which will not allow citizens of the United States to build and use ships as cheaply as foreigners, are removed; and if they are removed, then bounties will be no longer necessary, for ships will be procurable without bounties. Take another view of the situation. Having discouraged foreign trade by exorbitant taxation, we are now asked to heap up more taxes in order to encourage the same trade by cheapening freights. Having protected domestic wools by almost prohibitory taxes on the import of competing wools—especially the product of South America,—it is now proposed to tax the American wool-grower in order to secure lower transportation rates on the foreign wools that he regards as especially competitive. In the case of the trade between the United States and the Argentine Republic it is the judgment of experts that any subsidy to be effective, must be as great at least as will suffice to compensate the American ship-owner who exports, for the losses he now experiences from his inability to obtain a return cargo (paying freights) of imports. Neither probably will it be pretended, in the face of the intense competition to effect sales in the world's markets, that the profit accruing from any enlarged exports on account of subsidies, will be equal to the subsidy payments; or in other words, although a comparatively few ship-owners may gain, the people will lose.

But will the rates of ocean freights be cheapened to American exporters by the granting of subsidies to American-built vessels? It is barely possible that they may be to a limited extent under certain circumstances; but such a result is not probable. The subsidized owner will take his subsidy and charge for his freight service the current rates. If he reduces rates, his competitors will do the same, and the cost of ocean transportation will remain relatively as before. There are no unoccupied ocean trade routes; none in which abundant instrumentalities for the carrying trade would not be promptly and voluntarily supplied, if there was a chance of making even a small profit out of such supply. The recourse to subsidies is, therefore, a mere temporizing policy, and does not reach the broad problem, how to prevent the transfer of our whole ocean mercantile service to foreigners.

Again, our whole experience, in common with the experience of other nations, in respect to the payment of subsidies as a method of encouraging ocean navigation, has been unfavorable. The Federal exchequer was opened for years, in order that this mode of developing our steam-marine might have a fair trial; and what were the results? Before the war the Government gave large subsidies to the Collins, Havre, Bremen, Pacific, and other lines; but these contributions had no effect in preventing the continual decay of our merchant marine, and in 1861 there were no American ocean steamers away from our own coasts.

After the war, or from 1867 to 1877, when there was no war or Confederate cruisers to interfere with the development of our commerce and the use of American ships, the United States gave still larger sums in the way of subsidies; in the aggregate, $4,750,000 to the Pacific Mail Steamship Co., and $1,812,000 to the line between the United States and Brazil. The subsidy system as an agency for restoring our commercial marine had, therefore, during this period of eleven years, as fair a trial as possible, and the results it worked out so far failed to accomplish what its advocates had in view, and were connected with such a disgraceful chapter of Federal legislation, that Congress, in accordance with an almost universal popular sentiment, put an end to the whole business.

The more recent commercial experience of the United States is instructive in even a greater degree on the subsidy question. Investigations instituted by the *New York Com-*

mercial Bulletin show that there are at present 29 lines, with an aggregate of 125 steamers, regularly engaged in traffic between the United States and the West Indies, Central and South America. This service has greatly improved and increased within the last six years, or since 1883, and is now entirely ample to accommodate all existing demands for transportation. And yet, notwithstanding the enlarged opportunities for trading which have been thus afforded to us, there has been a decrease, in the same time, of our exports to the countries specified, to the extent of 12½ per cent. Our facilities for trading, so far as the carrying service is concerned, are especially efficient as respects the West Indies, and the ports of Mexico and Central America; and yet our exports to the West Indies were $5,316,848 less in 1888 than they were in 1883; and to Mexico and Central America $2,934,000 less. Going further south we find two lines of steamers subsidized by the Brazilian Government and making regular connection with our Atlantic ports; but the exports of our industrial products to Brazil were 21½ per cent. less in 1888 than they were in 1883. The main argument of the advocates of shipping subsidies, that we only lack means of transportation to secure a fairer share of the trade of the countries south of us, is clearly, therefore, by this record of experience completely demolished.

But while our exports to these southern countries have notably diminished, our imports from them have in the same time greatly increased—that is, we buy more of and sell less to our neighbors than formerly. To some this latter fact may seem paradoxical and inconsistent with the position before taken, that trade between the United States and Central and South America is restricted, because this country refuses reciprocal exchange of products. The explanation is, however, very simple, and there is no inconsistency of statement. We buy of the West Indies, Central and South America great quantities of sugar, coffee, tropical fruits, india-rubber, and hides, because we cannot produce these commodities at all, or in sufficient quantity to meet our demands for consumption, and can buy them cheaper of our southern neighbors than elsewhere. Under our existing tariff, however, especially our tariff on crude materials entering into our manufactures, we are not able to sell in the way of payment to the producers of sugar, coffee, fruits, rubber, and hides, such products of our industries as we would like to dispose of, and which the foreigner desires and needs to have ; because the latter can buy them cheaper in other countries—mainly in Great Britain. Our indebtedness to South America for our increasing exports is, therefore, settled in another way. England pays the bill in the first instance by the export of her manufactured products, and we pay England by the export to her of our agricultural products, cotton, cereals, meats, and the like ; and as a matter of fact, this roundabout, unnatural commerce represents what may be termed a triangular voyage. Thus, a ship loads at Rio Janeiro, for example, with coffee for New York, unloads there and reloads with grain for England, unloads there again, and again reloads with English manufactures for Brazil. And as English vessels are cheaper than any which Americans can build or have a right to own, the English vessels mainly do the work and earn all the profits, and English bankers and capitalists gather in the commissions and accruing interest on the capital employed. Or, to put the case more briefly, our existing tariff gives every foreigner every possible advantage for buying manufactured products in every market of the world other than our own.

DOES GREAT BRITAIN ENCOURAGE HER MERCHANT MARINE BY THE PAYMENT OF SUBSIDIES?

Prominent among the arguments brought forward in support of the proposition to attempt to arrest the decay and restore the prosperity of our merchant marine by

means of subsidies, or extraordinary payments on the part
of the Government, is the assertion *that the systematic
appropriation of large sums for the special object of encouraging
ship-using and ship-building has always been the practice and
policy of Great Britain;* and further, that to the continuance and present maintenance of such a system, is to
be attributed the continual advance and present great development of the British shipping interest. So frequently
and so unqualifiedly, moreover, have these assertions been
made in recent years, on the floor of Congress, by public
officials, by Chambers of Commerce, and by leading journals, and so seldom have they been questioned, that the
people of the United States have very generally come to
regard them as matters of history and of record which could
not be doubted. And the premises being at once accepted,
the conclusion was legitimate, that for the Federal Government to adopt the subsidy system was but to follow a policy which the long experience of the greatest maritime
nation had taken out of the domain of theory, and proved
to be eminently wise, practicable, and successful. All these
assertions, however, will be found on examination to rest
on no truthful or substantial basis; and to be what may
be properly designated as *historic* lies, originating mainly,
in the first instance, without intent to deceive, through
an imperfect understanding of the subject, and subsequently repeated and given credence on the basis of some
personal or supposed authority, without any attempt to in-
quire further as to their accuracy.

In support of this averment attention is asked to the
following statement of facts: The Empire of Great Britain
extends around the globe, and of its population of over
300,000,000, only about one eighth live within the territory
of the United Kingdom. To keep up a constant and regular communication with her detached colonies, military and
naval stations, by means of ocean sea-service, is a necessity

for the maintenance of the empire; just as much as quick and cheap communication with all of the States of the Union, by railway service, is necessary for the proper administration of the Federal Government at Washington.

It is not·to be denied that Great Britain, in maintaining this service, has expended, and is still expending very considerable sums; but it is important in this connection to recognize two facts. *First.* That in the days prior to 1860, when the sea service which Great Britain required was performed mainly by sailing vessels, she paid more than double per annum what she now pays for like service. But no one ever thought of regarding such payments, in the days of sailing vessels, as in the nature of subsidies for the encouragement of commerce and ship-building; and if they were subsidies, their influence did not drive the Americans from the ocean, or have any marked effect in expanding British shipping. *Second.* Previous to 1860 Great Britain paid as much as $5,000,000 in a single year for the transportation of her mails to and from the mother-country and its colonies, and foreign ports and dependencies. For the year ending March 31, 1889, the British Post-office Department, according to its report presented to Parliament, expended in all, for "conveyance" by land and by water, and by all agencies, the sum of £1,916,691; and as it is under this head that the so-called and much-talked of English steamship subsidies must be found, if found at all, an analysis of the items of such expenditure is of the first importance. And, instituting such an analysis, it appears that out of the above aggregate £903,634 was paid to British railway companies, and £637,502 ($3,100,-859) to steamship lines for mail conveyance; but of the latter sum, the "foreign market service" of steamships received £516,173 (or $2,508,590). If there was any thing in the nature of subsidy in this expenditure, it is clear, therefore, that the railways received the major portion; and that the comparatively pitiful sum of some three millions of dollars is all that the friends of subsidies can legitimately claim that Great Britain expended in 1888–89 for the support and encouragement of her immense ocean mercantile marine.

A word next in reference to certain expenditures by the British Admiralty, which are occasionally and somewhat mysteriously referred to as in the nature of important gratuities for the encouragement of British shipping. The simple facts in this case are as follows : The British Admiralty, independent of the Post-office Department, and without reference to any conveyance of mails, has paid, in recent years, under the head of war expenditures, comparatively small sums on account of certain steamships, on condition that they should be so constructed as to permit the carriage of heavy guns, and be made otherwise available as war cruisers, and thus modified be held at the disposal of the Government at all times, for purchase or hire, at the option of the Admiralty. The primary cost of such vessels being thus considerably increased, and their modified construction being also antagonistic to their most profitable employment in passenger or freight service, the British Government, of necessity, is obliged to make compensation for such losses. But upon what close calculations such compensation is rendered is made evident by the fact that, for the year 1888 the total expenditure for such purpose was only £22,380, while for the year 1889 an expenditure of £39,410 was estimated. For the year 1885 the amount thus expended was much larger, namely, about £600,000.

On the other hand the Post-office Department of the United States expended for the conveyance of the mails— mainly by land—in 1887–88, the large sum of $31,456,000, or more than seven times as much as Great Britain pays for a like service ; but no one pretends that this great expenditure was a subsidy paid for encouraging the building and use of American railway tracks, bridges, cars or locomotives, and yet it was a subsidy to our railroads in exactly

the same sense as the much smaller similar expenditure of
Great Britain was a subsidy to her shipping.

How the British government, moreover, entirely subor-
dinates whatever payments it may make for its ocean
marine service to the interests of the empire—its colonies,
its foreign dependencies, and military and naval stations—
rather than to its trade interests, is strikingly illustrated by
the way in which such payments are distributed. Thus
England's trade with Europe is very much greater than her
trade with Asia ; but she paid to the lines of steamers running to ports in Asia in the year 1888–89 for the conveyance of her mails £435,800, while to the lines plying
between England and European ports she paid during the
same year but £17,700, and this latter payment was entirely confined to the channel steamers running between
Dover and Calais, and Dover and Ostend. The simple and
truthful explanation of this is, that there are English colonies and military and naval stations in Asia, but none in
continental Europe. Again, the United States and the
West Indies are two countries about equally separated from
England. With the former England's trade is seventy-five
times greater than with the latter ; but the steamers performing service between England and the West Indies in
the year 1888–89 were paid £90,550, while those carrying the
mails between England and the United States during the
same year were paid but £85,000. If any man, after a
comparison of these figures, can wrest from them an interpretation that England's motive in paying thus extrava-
gantly for the transport of her West India mails, was to
build up her ocean marine, rather than maintain the in-
tegrity of her empire, and keep up regular and efficient communication with her colonies, he will be entitled to extraordinary credit for ability to manipulate figures in such a
way as to deduce from them any conclusion antagonistic to
the truth that he may think expedient. The precise object

2

which the Government of the United States has had in view in connection with its liberal grants of money and land to the great transcontinental railway lines—namely, to knit the widely separated portions of its dominion more closely together—has been aimed at by the British Government in its large contributions during the last forty or five-and-forty years to its various mail steamship lines which have united the colonies as never before with the mother-country. The nature of the service, both to the East and West Indies, has always been peculiar and exceptional, and still continues to be so ; and more than three fourths of the whole cost of the present ocean marine service is expended upon those two routes.

And here we find an explanation of a recent circumstance—namely, the reported grant by the British Government of £60,000 per annum to a steamship line, to run in connection with the Canadian Pacific R.R. from Vancouver to Hong Kong, which has been regarded as conclusive evidence that Great Britain builds up her ocean marine by subsidies. The Canadian Pacific, however, was built primarily, not for traffic purposes, but as a political necessity, to bind together the widely separate provinces of the Dominion of Canada ; and immense contributions of land and money were made by the Colonial Government to effect its construction. Once completed it opened a new, cheap, and expeditious route to the East, which England could control and use for the transportation of troops and 'munitions of war, as well as for postal service to India and Australia, in case European or Egyptian complications, which are always threatening, should close to her the Suez Canal. Her encouragement to the new line of steamers in question was, therefore, clearly dictated by military and not by mercantile considerations.

Again, it has been proposed during this present year (1889) to establish a new transatlantic fast line between

England and Canada, on the basis of a government bonus of half a million of dollars per annum. Such a compensation seems very large, and as having clearly for its object the encouragement of British shipping, but an examination of details showed that the proposed new line was not intended to be a freight-carrying line, but to carry passengers and mails primarily and almost exclusively; that the bonus was to be no assistance in moving British and Canadian products to a market; and, finally, that large as the bonus was, it was wholly insufficient to support a passenger line, pure and simple. The enterprise in question, therefore has, at least for the present, been abandoned.

It is not to be overlooked in this connection that certain of the British colonial governments do make compensation to certain ocean steamship lines independent of the home government; but this is done for the purpose of obtaining greater facilities for mail conveyance and for immigration, and not for the purpose of developing any shipping interest. Such compensation, for example, has been paid by New Zealand to steamships owned by Mr. Spreckles, an American citizen; and another line of American steamers receives payment from Brazil.

No little of confusion and misapprehension has attended the discussion of this subject, by reason of the different usage and signification of the term "subsidy" in the United States and Great Britain. In the former the terms "*subsidy*" and "*bounty*" are used as having an equivalent meaning; and when a subsidy is proposed it is generally understood to be in the nature of a bounty for the purpose of helping the owners of steamships to make a living, or earn profits. In the latter no payments are made by the Government to any steamships—other than the comparatively trifling Admiralty subventions above noted—except for the conveyance of its mails—an obligation as legitimate and as incumbent upon all nations desirous of maintaining correspondence with foreign countries, as are payments for the performance of similar service by railroads and other instrumentalities on land. But such payments in Great Britain, although spoken of as subsidies, are not bounties, or regarded as such by her Government or her people.

Great Britain, furthermore, pays no more to her ships than a fair commercial price for the service they render ; and the fact that all contracts for such service are always made after public advertisement and public competitive tenders on the part of all persons, native or foreign, who may desire to participate in the service, excludes the possibility of there being any thing in the nature of a benefaction or bounty, which could alone be authorized by direct and specific enactment of Parliament.

The statement that is also constantly made, that because the subsidies paid by England to English steamships enable them to carry English-manufactured commodities cheaply to her dependencies and foreign nations, therefore mercantile competition with England on the part of the United States in like business is impossible, is equally destitute of foundation.

In 1888 Great Britain owned seven twelfths of the world's shipping, and 70 per cent. of the world's steam-tonnage ; but out of this immense aggregate, not 2 per cent. performs any direct service for the British Government, or receives one farthing per annum from its Treasury in the way of payment for any thing. And yet the advocates of the subsidy policy in this country would have the American people believe that it is the employment of this small fraction of her marine tonnage by the British Government for mail service, and on the compensation for which not more than an average of 5 per cent. profit is probably realized, that makes Great Britain mistress of the seas, and gives her manufacturers advantages over American competitors in dealing with foreign countries.

Up to about 1850–51, the problem whether any ocean steamship could be navigated at a profit was a doubtful one. But in 1851 all doubt on the subject having been removed, Mr. John Inman, an English capitalist and merchant, possessing no more information or facilities than were available to other competitors, started his line of transatlantic screw steamers, which were to carry general cargoes and emigrant passengers, and be independent in every respect of the British Admiralty or Post-office. And from that time to this there has been a constant succession of other lines put in operation which have been pre-eminently successful, and which have never received Government aid of any kind,—not even compensation for ocean postal service. And these facts, which cannot be questioned or denied, also conclusively demonstrate the unsoundness of the assertion on the one hand, that the present great development and supremacy of British ocean navigation is due to the continued payment of subsidies by the Government; and on the other, that Government aid in the way of subsidies is, and has been, necessary for the resuscitation of the American mercantile marine, unless it is at the same time assumed that the Americans are an inferior race, and are unable to do under equal circumstances what the Englishman has found no difficulty in accomplishing. And if circumstances have not been equal, it is because our navigation laws and fiscal policy would not permit it.

A further point of importance should also not be overlooked by those desirous of getting at the truth of this matter. The sailing fleet of Great Britain is the largest in the world, and in 1888 numbered 15,025 vessels, representing over three millions of tons; but not one of these vessels is employed by the British Government. In general, however, they engaged in profitable ocean service, while our sailing vessels are rapidly decreasing in number because they are unprofitable. And yet no one can deny that the same opportunities of freight in the general ocean-carrying trade are open to British and American sailing ships, excepting that the latter have the advantage of being protected in their coastwise trade, while the coasting trade of Great Britain is open to all nations. And this statement alone ought to be convincing that the British carrying trade on the ocean is not maintained and made prosperous by subsidies.

The utter want of all similarity between the ocean service which private-owned steamships render to the British Government and the object for which it is proposed to pay subsidies to shipping in this country should not be overlooked. In the one case payments are made for service, based on contracts awarded after public competition. In the other a subsidy is to be given on the basis of the mileage sailed. In the former the prime object proposed for attainment is the carrying of letters ; in the other the sailing of the ship or the carrying of the flag. There is something very sentimental and captivating in the assertion that " trade follows the flag." But trade does nothing of the kind. It follows the dollar wherever it is to be found, and in the attainment of this object the question of the flag to those concerned in the trade is a matter of very little consideration. Goods seeking transportation will never wait long upon a dock, because the vessel moored to its side and ready and capable of transporting them carries a foreign flag.

Finally, when England's record in this matter is examined, it becomes apparent that her so-called subsidy policy has no characteristics antagonistic to the principles that underlie and govern all correct and shrewd business transactions. She 'subsidizes ships in the same sense as the citizen subsidizes the butcher, the baker, the grocer, and the dry-goods merchant ; that is, she avails herself of the

THE DECAY OF OUR OCEAN MERCANTILE MARINE.

services of a very small proportion of her ships and ship-owners for carrying her mails and pays them for it in exactly the same way as the United States pays railroad, steamboat, and stage owners for performing similar service. And in all her history Great Britain has never appropriated a dollar for the purpose of aiding in the construction and employment of a British merchant-ship, and no person can point to a single act of Parliament that ever gave a bounty or subsidy for such purpose. The testimony of all British authorities runs to the same effect. Thus, in 1881, the late Mr. Henry Fawcett, M.P., then British Postmaster-General, declared explicitly that " a postal subsidy is simply a payment made for the conveyance, under certain specified conditions as to time and speed, of postal matter"; that " such subsidies are not granted with the object of giving to English shipping any protection against the competition of the shipping of foreign countries," and mentioned as proof of the correctness of this assertion, " that when a contract for the conveyance of mails is advertised, no restriction whatever is imposed upon any foreign vessels competing," and that " the subsidy would be paid to foreign-owned and foreign-built vessels if it was considered that the best and cheapest service could be thus secured."

The largest amount specifically paid by the British Government for ocean service is to the so-called " Peninsular and Oriental Steamship Company," which carries the mails, Government despatches and messengers between England and the East, and the receipts and experience of this company are often cited as evidence that England not only called it into existence, but has always maintained it by the payment of subsidies, in the American acceptation of the term. But on this point Mr. W. H. Lindsay, the leading authority on English shipping, speaks thus decisively :

" The impression," he says, " that this company owed its origin to Government grants, and that it has been maintained by subsidies, is not supported by facts. Whether the company would have continued to maintain its career of prosperity without Government subsidies, is a problem too speculative for me to solve. Free from the conditions required by Government, the company would probably have done better for its shareholders had it been also at liberty to build and sail its ships ~~ it pleased, despatching them on such voyages and at such rates of speed as paid it best ; and in support of this opinion I may remark that various other shipping companies, with no assistance whatever from Government, have yielded far larger dividends than the Peninsular and Oriental Company ; and further, that private shipowners who never had a mail bag in their steamers have realized large fortunes." And again, commenting on the large payments made to this line by the Government, he says : " From whatever cause it may have arisen, the fact is apparent that, though the annual gross receipts of the company are enormous, its expenditure is so great that less balance is left for the shareholders than is usually divided among those undertakings of a similar character which receive no assistance from Government, but are free to employ their ships in whatever branch of commerce they can be most profitably employed."—*Lindsay's Merchant Shipping.*

The late Mr. Guion, founder of the Williams & Guion line of steamers, has also placed himself on record, that his company " never received a penny of Government subsidy and felt no necessity for it."

Within the last year the British Post-office authorities have made a contract with the North German Lloyd for a regular mail service between Southampton and New York, in preference to employing the Cunard and White Star lines, for the reason that the Government could effect a saving under the new arrangements to the reported extent of £25,000 per annum. Commenting on the change, Mr. John Burns, of the Cunard Company, in a recent communication to the London press made the following statements :

" Whatever the saving made by the employment of the North German Lloyd ships may be, the acceptance by this company of

lower rates than English companies is accounted for by the following considerations : 1. They enjoy a large subsidy from their
own Government, an advantage denied to British ships. 2. They
are not subject to the restrictions and regulations of British law.
3. They have not to call and wait at Queenstown. 4. They call
at Southampton on their way from Bremen to New York in order
to compete for British traffic. Whatever they can obtain for the
carriage of the mails is therefore practically so much clear gain,
and helps them in their war on British trade."

So according to Mr. Burns, who must be recognized as
authority, the British Government at the present time (whatever it may have been before), in place of encouraging, is at
war with British trade.

In short, all this attributing the maritime prosperity of
England to subsidies is a concealment of a truth that it is of
the utmost importance for the American people to learn,
namely, that England is first in shipping, because she is
first in commerce ; and she is first in commerce, because she
has freed her trade and her ships, while the United States
have shackled the one and destroyed the other.

So much then for the experience of Great Britain in
respect to her so-called subsidy policy. On the other hand,
there is no doubt that certain of the continental nations of
Europe have in recent years attempted to stimulate ship-
building and ship-using by a carefully devised system of
subsidies, in the nature of bounties, the same substantially
as it is now proposed shall be adopted by the United States.
The results of these experiments have thus far been complete failures ; and as France has taken the lead in this
policy, and most thoroughly carried it out, attention is
especially asked to the following record of her experience.

In 1881, the French Government offered to give a bounty
of twelve dollars a ton on all ships built, in French yards,
of iron and steel; and a subsidy of thirty cents per ton
for every thousand miles sailed by French vessels ; and as

they did not desire to put any inhibition on the citizens of France buying vessels in foreign countries and making them French property, in case they desired to do so, they proposed to give one half the latter subsidy to vessels of foreign construction bought by citizens of France and transferred to the French flag.

At the outset, the scheme worked admirably. New and expensive steamship lines were organized with almost feverish haste, and the construction of many new and large steamers was promptly commenced and rapidly pushed forward in various French ports, and also in the ship-yards of Great Britain and other countries. The Government paid out a large amount of money, and it got the ships. In two years their tonnage increased from a little over 300,000 to nearly 700,000 tons for steamers alone ; while the tonnage engaged on long voyages increased in a single year from 3,600,000 to over 4,700,000 tons.

It was probably a little galling to the French to find out after two years' experience that most of the subsidies paid by the Government were earned by some two hundred iron steamers and sailers, and that over six tenths of these were built and probably owned in large part in Great Britain ; so that the ship-yards on the Clyde got the lion's share of the money. But as all the vessels were transferred to and sailed under the French flag, and were regarded as belonging to the French mercantile marine, every thing seemed to indicate that the new scheme was working very well, and that the Government had really succeeded in building up the shipping of France. But the trouble was that the scheme did not continue to work. The French soon learned by experience the truth of the economic maxim that ships are the children and not the parents of commerce ; and that while it was easy to buy ships out of money raised by taxation, the mere fact of the ownership of two or three hundred more ships did no more to increase trade than the purchase and ownership of two or three hundred more plows necessarily increased to a farmer the amount of arable land to plow ; or, in other words, the French found that they had gone to large expense to buy a new and costly set of tools, and then had no use for them.

And, what was worse, they found, furthermore, that while they had not increased trade to any material extent, they had increased the competition for transacting what trade they already possessed. The result has been that many French shipping companies that before the subsidy system were able to pay dividends are no longer able ; fortunes that had been derived from the previous artificial prosperity have melted away, and the French mercantile marine ceased to grow—only $601,120 being paid out for construction bounties in 1886, as compared with a disbursement of $908,000 in 1882. In fact, the whole scheme proved so disastrous a failure that the late Paul Bert, the eminent French legislator and orator, in a speech in the French Assembly, seriously undertook to defend the French war of invasion in Tonquin on the ground that its continuance would afford employment for the new French mercantile marine, which otherwise, we have a right to infer, in his opinion would have remained idle. A recent writer—M. Raffalovich—in the *Journal des Economistes* has also thus summed up the situation. " It may be asserted," he says, that " the bounty system in France, which was intended to bridge over a temporary depression, has aggravated the situation, and has proved itself to be a source of mischief, not of cure."

The experience of the mercantile marines of Europe also affords the following curious results during the eight years prior to 1880 and before the inauguration of the French bounty system. French shipping, in its most valuable branch-steamers, increased faster than the shipping of any of its Continental competitors; but after 1880, the increase in the steam marine of Germany, where no bounties were paid, was relatively greater both in number and tonnage of vessels than in France where large bounties were given after 1881; and was also greater as respects the aggregate tonnage of all vessels—sail and steam. The obvious expectation of the French Government in resorting to the bounty system for shipping was that ships built and navigated with the aid of the bounties would carry French manufactures into foreign countries, and thus open new markets for domestic products. But experience, thus far, has shown that all that has been effected is a transfer, to some extent, of the carriage of goods formerly brought in foreign vessels to French vessels; while, on the other hand, the increase of tonnage, under the stimulus of the bounties beyond the requirements of traffic and the consequent reduction of freights, has entailed "a loss, and not a gain, to the French nation, by throwing upon it the burden of a shipping interest that, but for the Government aid, would have been unprofitable, and which, because of such aid, can not conform itself to the demands of trade."

The experience of Austria-Hungary in attempting to find new outlets for their produce, or fresh employment for their shipping by the payment of subsidies, has been analogous to that of France, and equally unfortunate. The steamers of the Austrian Lloyd Company have made more voyages to the "Far East" than when unsubsidized; but the exports of Austrian products have not materially increased, while the mercantile marine generally of Austria is rapidly declining.

Contrast these results with the experience that has accompanied the free ships and free commercial policy of Great Britain. Of the total increase in the shipping trade of the principal maritime nations from 1878 to 1887, one third occurred in British tonnage; while of the increase in the merchant steam tonnage of different countries, during the same period, nearly two thirds is to be credited to Great

Britain. In the year 1887 the mercantile navy of Great Britain, while carrying three fourths of the whole of her own immense commerce, carried at the same time one half of that of the United States, Portugal, and Holland ; nearly one half of that of Italy and Russia; and more than one third of that of France and Germany. As the ocean mercantile tonnage of the United States declined between the years 1878 and 1887 in a greater degree than that of any other country, it is very clear at whose expense the increase in the shipping of other nations was made during this same period.

REMARKS OF MR. WILLIAM J. COOMBS.

I have been invited to follow the eloquent paper which has been read to us this evening, by a few extemporaneous remarks. I have consented for the reason that I quite well understand that your object in inviting me to speak, is to get the merchants' view of the question, and to come into possession of such facts in my experience as may bear upon the situation.

It is undoubtedly a subject for deep mortification that a great nation like ours, which at one time was able to dispute the dominion of the seas with the strongest maritime powers, should at this time of general prosperity be without a mercantile marine. If we look for the cause of this condition, we shall not find it in the poverty of the country or in its lack of resources, but in unwise restrictive legislation, which has made it impossible for us to avail ourselves of the splendid resources that nature and the enterprise of our people have put at our disposal. I shall not attempt to argue this evening, whether or not this legislation was wise and proper at the date of its conception ; there may easily be a difference of opinion upon this point, and it is not worth our while to waste time upon it,—what we have to do with, is the present and the future. If we find that these laws were enacted to meet requirements and conditions which no longer exist, and that they fetter us now under our new conditions, and make it impossible for us to keep pace with the nations of the earth, so that we are daily losing our supremacy, it becomes our duty not only to ask, but to demand their repeal.

In my opinion, the question of free ships is at present a matter of secondary importance,—there are questions antecedent to that which should claim our attention, and the settlement of which, if properly made in the interests of commerce and of the general good, will carry with it the cure of all these minor errors. I believe that if we to-day had a mercantile marine given to us, free of cost, we could not sustain it and make it profitable. In order to run ships, either sail or steam, at a profit, there are two things that are certainly necessary, viz., outward cargoes and return cargoes. I claim and can prove from my experience, that we have settled the first half of this problem, but that the second half is unsettled, and is the cause of our present embarrassment. Not only does it make it impossible for us to compete for the carrying trade of the world, but it prevents our shipping, upon any reasonable terms, the products of our factories and fields.

The time has come when the enterprise of our manufacturers, combined with the intelligent skill and inventive genius of our mechanics, enables us to compete in the markets of the world with the European factories, as regards nearly every class of our industrial productions. The exceptions are so few, that they can be easily enumerated, and consist in most part of those articles in which the cost of the raw material constitutes the most important item. We compete most successfully in the classes of goods in which the item of labor is the largest factor. However, we find ourselves in the position of the farmer who has tilled his fields and raised his crop, but who has failed to provide wagons in which to carry it to its destination.

The thing which now gives the merchant who sells to foreign markets the greatest anxiety, is to procure vessels, either American or foreign, at reasonable rates of charter, to carry the goods to his customer after they are sold. In this particular, he is at a great disadvantage as regards his European competitor. The owner of a vessel who charters it for a voyage to Buenos Ayres or the Cape colonies, does so with no expectation of getting a return cargo to this port, but calculates upon taking one from there to Europe. If he wishes to return to this country, he must, except under unusual circumstances, come back in ballast. His rate of charter is fixed upon this basis. As an example I will state that to-day we received a letter from the captain of a vessel which we loaded for Natal, Africa,—informing us of his intention to proceed in ballast to Brazil to take cargo for New York.

Our house has repeatedly within the present year, chartered vessels in foreign ports to come to this country in ballast, in order to take away merchandise for which we had orders. At the present moment we have six of such charters pending. This not only involves long delays but enormously increases the cost ; and for both reasons, puts us at a great disadvantage in comparison with our European rivals. This is a state of things which the law of supply and demand would remedy, were it not for our unwise restrictive legislation, by which we are prevented from receiving certain classes of raw material which our manufacturers need, and which would furnish excellent return cargoes for our vessels.

There is a proposition pending,—which will without doubt be pressed before the Congress soon to convene, to remedy this by a system of subsidies and bounties, but I venture to assert that unless our government is prepared to grant assistance to at least the approximate amount of the freight of a return cargo, it will not have the desired effect. It will open the way for a vast amount of jobbery and corruption, and at the best will only be a temporary expedient, a correction of one error by the commission of another. Government can very materially and legitimately assist any proposed steamship line by paying liberally for the transportation of mails. Those lines already established have had just cause for complaint on account of insufficient remuneration. Such assistance is all the more proper, for the reason that carrying the mails is one of the things which by general consent is left for the Government to do.

During my canvass for Congress in 1888, I often pointed out what I considered to be the first step in the right direction, to-wit: the removal of the almost prohibitory duty on South American and Cape wool. This would give our woollen manufacturers what they sorely need, and, I verily believe, would put them in a position to compete with the European manufacturers in the markets of the world—which they are now unable to do. During the election of 1888 they resisted the attempt, but already wiser counsels begin to prevail, and the time is not far distant when the demand will become imperative, and will have to be conceded, if it is not complicated by other claims. It will, I am sure, be found that such action will not injure our own wool-growers, but will result in a more active market for all the wool that we can produce under favorable circumstances. If the duty is removed I believe that, within two years, any attempt to impose it would be met with the same derision which was encountered by a similar proposition in relation to hides.

I might go on and enumerate other raw materials from which the duty should be removed, but I believe that it is good policy to leave them, for the present, out of the discussion. I think that a very serious error has been made by revenue reformers in attempting too much at a time, and thus banding all the selfish interests in a common defence. If we can succeed in breaking their forces by securing the adoption of a proper policy in relation to one important item, we shall soon have other things falling into line. For that reason I would not disturb the present duties on manufactured goods, except in cases where protection has led to the formation of trusts and combinations against the public interests. Already they have, in a great variety of cases, become like the weeds in the bottom of the channel: the tide has risen above them. Except in the cases mentioned, they will not materially interfere with our prosperity. The great competition, engendered by success, has had its natural result in overproduction, thus giving back to the people the benefit of reasonable prices.

The scales are now very evenly balanced between our own and the European manufacturers, with a decided tendency to dip in our direction. If we can successfully compete even to a limited extent, embarrassed as we are by duties on our raw materials and by dear transportation, what could we not do, with free raw materials and proper facilities for the cheap delivery of our goods? With these given to us by the repeal of unwise legislation, our manufacturers could successfully compete for the trade of the world; they could vastly increase the capacities of their factories; skilled workmen would be in demand, at remunerative wages, and would in a measure replace the foreign and unskilled laborers who are now employed in our mines in providing the raw materials,—thus the average quality of our citizenship would be raised. The agricultural interests of the country, whose foreign markets are now seriously threatened by new competitors in the raising of grain, would find a better and more remunerative home market. This is not a fanciful picture; on the contrary, the result predicted is clearly within our reach.

I cannot prolong my remarks indefinitely, as I have already trespassed too far upon your time; I will detain you only enough to urge upon you—and through you upon every good citizen—not to delay action too long.

We are accustomed to look upon the prosperity of our country as unassailable, to imagine that we have in some way a patent upon success. We must not be deluded by this idea. The time was, when we were the foremost nation in rapid progress. If we look around, we will see that the whole world has awakened to activity. Fresh competitors are springing up in all directions, and nations that have been asleep have risen to a new life. Under the circumstances, to stand still is relatively to go backwards; to adhere to old and crude methods at the time of a general advance, is suicidal. A selfish and grasping policy, by which we seek to secure every thing without giving any thing in return, should debar us from the sympathies, as it surely will from the general prosperity, of the world.

www.ingramcontent.com/pod-product-compliance
Lightning Source LLC
Chambersburg PA
CBHW022025190326

41519CB00010B/1600